Science
Fair
Success

How to Excel in Science Competitions
Revised and Updated

Melanie Jacobs Krieger

Enslow Publishers, Inc.
40 Industrial Road   PO Box 38
Box 398   Aldershot
Berkeley Heights, NJ 07922   Hants GU12 6BP
USA   UK
http://www.enslow.com

Library of Congress Cataloging-in-Publication Data

Krieger, Melanie Jacobs.
     How to excel in science competitions / Melanie Jacobs Krieger.
  — Rev. and updated ed.
       p. cm. — (Science fair success)
     Includes bibliographical references and index.
     Summary: A guide for the high school student researching a
science project for entry in a competition.
     ISBN 0-7660-1292-1
     1. Science projects. 2. High school students—Competitions. 3. Research—
Competitions. [1. Science Projects. 2. Research—Competitions.] I. Title. II. Series.
Q182.3.K75   1999
507.8—dc21   98-31754
                                                       CIP
                                                       AC

This book is a revised and updated version of
How to Excel in Science Competitions published in 1991.

Printed in the United States of America

10 9 8 7 6 5 4 3 2 1

To Our Readers:
All Internet addresses in this book were active and appropriate when we went to press. Any
comments or suggestions can be sent by e-mail to Comments@enslow.com or to the address on
the back cover.

Illustration Credits: © Corel Corporation, pp. 5, 13, 23, 31, 43, 57, 65, 75; Melanie
Jacobs Krieger, pp. 6, 9, 11, 15, 17, 24, 28, 29, 32, 35, 36, 39, 46, 48, 50, 54, 59, 61,
63, 67, 72, 76.

Cover Illustration: © TSM/Chuck Savage

Contents

To my mother and father,
Isabelle and Leo Jacobs,
who always asked,
"Did you mention me?"

Chapter 1

Why Compete?

Science research projects are quite different from the projects one usually finds at a junior or senior high school science fair. These investigations are not simple models, a collection of objects, or a report, but an independent research experiment created by the student using the scientific method.

An independent science research project can be a very demanding undertaking that involves a great deal of work, time, and effort. Because your project will be on a topic that you select and enjoy, your experience should also be fun and rewarding.

Christine was an excellent high school student. She loved school, especially biology. But her high school did not offer any opportunities to study biology other than regular class work, lectures, and simple labs. Christine's dream was to work in a biology research lab and eventually carry out some of her own experiments and investigations. She became an intern in a genetics lab at a nearby university. Her principal tasks were washing and cleaning glass instruments, but she loved the work because she was becoming familiar with the laboratory routine and learning many research skills.

She became so committed to the work that she even persuaded her mother to give up her lunch break to drive her to the lab so she could continue her work during the school year and during school vacations. Christine continued her work during the summer of her junior year. However, during that time she never worked on a research project of her own.

In the fall of her senior year, Christine returned to the lab and asked whether there was a research project she could do by herself to enter in the Science Talent Search. The Science Talent Search was formerly sponsored by the Westinghouse Corporation and is now sponsored by the Intel Corporation. Christine decided to work on a genetics project in which she

Working in a laboratory helps you learn scientific technique.

tried to verify the theory that you can predict the occurrence of Down syndrome by studying a person's genetic code. Christine studied the nuclei of certain cells to try to understand why some chromosomes did not split during cell division. Christine worked furiously on her project but encountered many problems with her silver nitrate staining solution. There seemed to be an impurity in the solution, so she tried ordering a new mixture, changing the utensils used to spoon out the mixture, and even slightly modifying her solution and procedure. But the deadline to write her paper approached. When she wrote her paper, titled "Double NOR (dNOR) as a Risk Factor for Trisomic Offspring," Christine described the problems that she had encountered and the ways she had devised to cope with her difficulties.

She was selected as a semifinalist in the Science Talent Search. Although her project never worked, Christine learned a great deal about genetics and was given advanced standing in her field of study at her chosen university. She also received a full tuition scholarship because of her scientific research.

Christine is just one of hundreds of high school students who experience the benefits of entering science research competitions. Any student who is willing to make a commitment to work on an independent research project can receive them.

You will learn how to conduct and complete your own independent science research project using the scientific method. You will learn how to formulate a hypothesis and the proper methods to use to prove or disprove that hypothesis.

Student researchers acquire a great deal of knowledge about a specific topic that is usually far outside the realm of the curriculum they would ordinarily study in high school.

Whatever you choose to study—mathematical chaos and fractals, motion of ocean waves, brine shrimp, nerve muscles, cell regeneration, statistics, or even political science—your knowledge will far outdistance the everyday curriculum.

As you work on your science research project, you will discover that the most important aspect of research is the exchange of ideas and discoveries. Therefore, you will need to learn how to write a research paper and abstract. If your research proves to be successful, your paper may even be published in a scientific journal.

Many high school students have been able to successfully publish articles about their research projects. For example, David worked as a student researcher for two summers in the anesthesiology department at a university hospital. While he was there he worked on many different experiments and was cited as a coauthor of "The Determination of the Mechanism of Anesthetic Action Through Electro-Neurophysiological Analysis of BC3h-1 Murine Intracranial Tumors" that was published in the journal *Anesthesiology*.

In addition, you will learn how to present your own work either by giving talks before a large audience in a lecture hall or by making poster presentations. You will find that you will be able to meet distinguished scientists and judges to explain and discuss your work.

As you work on your independent research project, you will become very proficient in your ability to use the library. A research scientist must be able to find and use books, periodicals, current scientific journals, as well as the Internet and computer databases to acquire the proper background for

During a poster presentation, a student's poster display should make an immediate impact on the judges.

experiments. You will learn how to use primary and secondary sources.

Science research projects and competitions can be financially rewarding. Many colleges offer merit scholarships to high school graduates who have worked on science research projects and entered major science competitions. These students receive internships in the university's laboratories. They use the skills and techniques they have acquired while working on their projects to continue their research during their undergraduate years. And, of course, many science research competitions award generous scholarships to the winners.

As you work on your science research project, you will discover that you are acquiring benefits that you did not initially expect. You will be pleasantly surprised to discover that professional scientists are usually happy and willing to talk to you about your topic and research. In fact, they will frequently invite you to visit and tour their research labs. They might even invite you to work in the labs, either on your own project or as part of a team working on a larger project.

You will receive equal responsibilities and privileges with those scientists and graduate students working in the research lab. Of course, you will be expected to hold your own, meet important deadlines, and work independently.

If you are working on a small part of a very large project, then your responsibilities will be even greater. The other scientists will be waiting to see the results of your project before the rest of the experiment can continue.

While working in the lab you will become familiar with many types of scientific equipment. At first, you may be no more than a glorified assistant. As you continue your work, however, you will discover the different names and uses of the various pieces of apparatus found in the lab. As you become more experienced, you will be able to prepare and set up some of the equipment for your own experiments.

Working on a science research project can actually help you determine the kind of work you could possibly do as an adult. Many of the skills you learn will stay with you throughout your college and professional career.

Student researchers have stronger college applications. As you begin to think about college applications and scholarships, you will discover that there are many high school students who are valedictorians, salutatorians, and in the top 5 percent of their graduating class. Many times, these competitive students have high SAT scores and have been active in many school and community activities. You will find, however, that there are

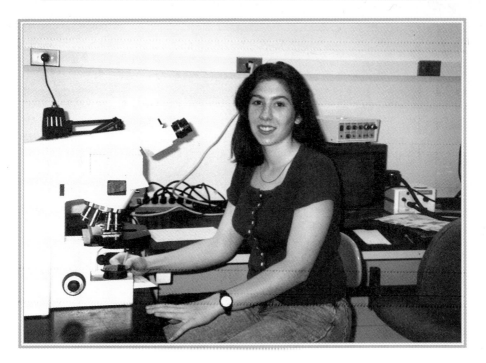

Research students often use state-of-the-art equipment.

not many students who have successfully completed independent research projects for contests such as the Intel Science Talent Search. In one recent year, although seventeen thousand students requested the application materials, only sixteen hundred students submitted completed applications and research projects.

A research project can give your college application essays a focus. You will want to share with other people the advanced knowledge and experiences you have gained. Whereas many students will have quite a bit of trouble answering college application essay questions such as "Describe an experience outside the classroom that has been meaningful to you," you will have many experiences from which to choose.

College interviews will also be a lot of fun. Whereas many other students will have difficulty finding topics to talk about, you will discover that you have your research experiments to share. You will have fun discussing your successes and failures, the students and professors you have met, your travel to other cities, and your lab experiments. In addition, you will have self-confidence while talking to your interviewers because of your experience in talking to judges at the various competitions.

However, there is more to the competition than entering for the sake of winning. To enter is to win. You will become quite a different student. You will have met very stimulating people, had many interesting experiences, and have become knowledgeable in a specific field. In addition you will have learned how to write and present a scientific paper. And most important, you will have discovered that just by doing your project, you have actually won before you have even entered!

Chapter 2

Profiles of Student Researchers

High school sophomore Jennifer knew immediately that she would like to work on a science research project. She talked with several biology teachers in her high school. One teacher suggested that she contact the biology department at the local college. Jennifer did, even though that telephone call was probably one of the hardest she had ever made. She saw the chairman of the department, and he said she could work in his department as long as she followed a strict working schedule and did not damage anything in the lab. Jennifer worked twice a week after school during her sophomore year and throughout the summer. During this time, she read about the experiments in her lab, learned lab procedure, and learned to use specialized equipment. She hopes to begin working with a graduate student on a project of her own in the spring of her junior year.

A High School Junior: Scott

Scott was a member of a high school science research class. He participated in many different projects and contests. Some of his activities included the Science Olympiad, a bridge building competition, a science fair, and a statewide energy research competition. But Scott's favorite activity was working in the lab on his independent research project in immunology. He enjoyed his experience immensely and is planning to write about his experiments for the Intel Science Talent Search. Scott summarizes his experience this way:

> More important than the project and its results, I found that I enjoyed the learning experience itself. Never would I have thought that something that could start out so tedious (three weeks of background reading) could turn out to be so interesting. I have learned more than just some immunology. I have learned a method of thinking, and this is perhaps the most valuable lesson I have learned at the lab.

Science Talent Search Entrant: David

David attended special summer programs in math and science each year in high school. He spent several summers at the Johns Hopkins Center for Academically Talented Youth (CTY), positions that are offered to students with a total SAT score of more than 1000 at thirteen years of age. David studied astrophysics, marine ecology, and astronomy. He then spent a summer at the Los Alamos Student Science Program for talented youth, studying optical systems.

During the summer of his junior year, David studied at the Research Science Institute (RSI), an institute devoted to helping students work on individual science research projects. Positions in the program are selected on a competitive basis. During his summer at RSI, David developed a set of simple and original experiments dealing with the theories of mathematical chaos. David finished his work with a paper titled "Toward a General Definition of Chaos." David was named a semifinalist in the Intel Science Talent Search.

Science Talent Search Entrant: Marc

Marc took a different approach to entering research competitions. He worked on several different aspects of one project for two years. He was interested in the power of ocean waves breaking in shallow water. In his sophomore year, he

Thomas Petersen was a finalist in the Science Talent Search. Thomas completed a project on polymers.

studied the effects of wave trains on the open ocean, and then in his junior year, he worked on the prospect of harnessing the energy from breaking waves. Marc then worked on a project to create a mathematical model of waves breaking in shallow water. For his Science Talent Search project, he continued with his interest in ocean waves and created a project titled "On the Trochoidal Waves of Gerstner and Rankine." Marc was named one of the three hundred semifinalists in the Science Talent Search.

Marc also worked on building musical instruments for the Sounds of Music contest in the Science Olympiad. He constructed a working pipe organ out of garden pipe, and a cello out of a plastic garbage can.

During the summer after his junior year, he worked on superconductors at the Brookhaven National Laboratories.

Science Talent Search Entrant: Sumit

Sumit worked on his project for two summers at the hematology labs at Stony Brook University on his project entitled "The Characterization of the Generation and Inhibition of Factor Xa in the Human Blood Coagulation System." He summarized his experience:

> The professors thought I was a grad student—and treated me that way. At first, I thought my job wasn't the greatest. It looked like my friends had wonderful jobs as busboys—they were making lots of money and I was not. But then they said that being a busboy got very, very boring. My research got more and more exciting. Incredibly, my experiments really worked. I was in an adult's world. I was treated with respect and it was just expected that I would do my work. I encountered lots of frustrations. Many of these frustrations were over very small things. For example, the pipettes didn't work correctly so all the work

involved with the pipettes was slowed down. Then the solution's labels were mixed up—or so we thought. And so, two weeks worth of work had to be thrown out!

I had the best summer of my life. I know exactly what it's like to be a research scientist. I was one for a summer. I liked the fact I had lots of free time some days. I would set up the experiment, get it going, and then be able to listen to music or meet my friends for a while. It was great. I loved it. Everyone should do it!

Sumit was named a semifinalist in the Science Talent Search.

Science Talent Search Entrant: David

David worked on many different science projects and one long-term project for different competitions while in high school. He worked on a project for his state's energy research

Clyde Law was a finalist in the Science Talent Search. He completed a project on nuclear matter.

competition dealing with exothermic chemical reactions. He used an inexpensive chemical reaction to create heat for homes and offices. This project earned David fourth place overall in the competition.

At the same time, David worked on a long-term neurophysiology project at Stony Brook University during the summers of his junior and senior years. David finished his project and wrote a research paper to enter in the Science Talent Search. Although he did not win the competition, his paper was selected to be presented at the Long Island Science and Engineering Fair.

Science Talent Search Entrant: Carl

Carl worked on one project consistently during high school. Carl became interested in discovering whether plant roots respond to centrifugal force. His final project was titled "Control of Growth Patterns: The Response of Bean Roots to Centrifugal Force and Gravity."

In his sophomore year, Carl worked on his project, performing some simple experiments and research. He also built a centrifuge consisting of a bicycle wheel driven by an electric motor that was mounted on a wooden frame to grow his plants.

Carl entered a regional science fair in his sophomore year. Although he did not win any prize or recognition for his work, he was undaunted. During the next year, he tried many different and more sophisticated experiments involving bean root growth, gravity, and centrifugal force. He did much reading and library research. He completed his work and again entered the regional fair. This time, the project won best of fair! He then went on to the state science fair competition. This time his work was chosen as best in the state. Carl made some minor modifications and entered the project in the Science Talent Search, the International Science and Engineering Fair, and the Junior Science and Humanities Symposium. Carl did not win recognition at any of these competitions. However, his perseverance and experimentation helped him get accepted into the Ivy League college of his choice.

Science Talent Search Entrant: Mathew

Mathew became interested in science research when he was in the ninth grade. At that time, his older brother entered the Science Talent Search with a project on experimenting with a type of white blood cell known as neutrophils. Mathew asked to do some work in the same lab as his brother at a local university. He performed many experiments and did extensive reading during his freshman year.

His project, "Neutrophil Movement," was entered in the regional science fair. Mathew won first place, even though he was only in the ninth grade and competing with students in all grades in the senior high school competition. The project was then entered in the state science fair competition. Mathew was declared the overall state winner.

Mathew continued with his research in the tenth grade. He entered the regional Science and Engineering Symposium. Mathew was chosen to speak and was selected as an alternate in the biology category.

In Mathew's junior year, he continued his research and entered the regional science fair. His project was declared best in the fair, and he was selected to enter the state science fair competition. Mathew became a finalist.

During the summer before his senior year, Matt continued work on his project and wrote a final research paper, "Movement of Neutrophils Stimulated by Chemoattractant" to be entered in the Science Talent Search. Matt became a semifinalist.

At his college interviews, Mathew met with various members of the biology faculty as well as the usual interviewer to find the college that would allow him to continue his research during his undergraduate years. He will continue work on his research in college.

Science Talent Search Entrant: Raphael

Raphael worked during the summer and spring of his junior year at the hematology labs at Stony Brook University. Hematology is the science dealing with blood. During that time he worked with different blood coagulants and studied the effects of different inhibitors of coagulation. The paper he wrote to describe his research experiments was entitled "Weak Inhibition of Activated Human Coagulation of Factor X by Soybean Trypsin Inhibitor (Kunitz)."

Raphael's research did not win the Science Talent Search or any other prize. However, while in his freshman year of college, he discovered that he could easily handle junior-level chemistry work because of all the lab work he had done, and he was placed accordingly.

Raphael's professors read his high school paper and were awed by its sophistication, and he was, therefore, awarded a summer position in the chemistry lab as a freshman. Usually

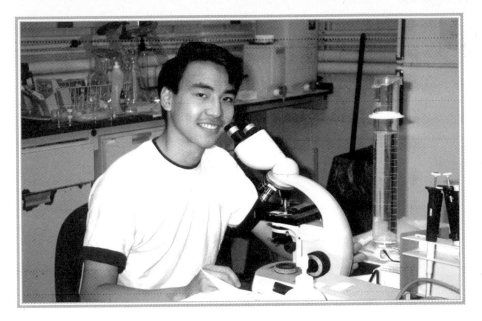

A student works on his research project in a lab at Stony Brook University in New York.

these positions are awarded to juniors or graduate students. During the summer, he worked alongside graduate students who were working on their research experiments for their Ph.D. degrees.

Science Talent Search Entrant: Amanda

Amanda was interested in entering the Science Talent Search with a project in psychology using statistical analysis techniques. She spent a great deal of time during her junior year on preliminary work. She studied statistics in high school, then she took a course in psychology at the local university. During the summer of her junior year, Amanda took a second course in psychology and formulated the plans for her project.

In her experiment, Amanda showed both a control and an experimental group of students the same set of ink blots similar to the well-known Rorschach plates. She asked each group of students to describe the ink blots. She then read different directions to both groups to determine whether or not the subliminal suggestions found in the experimental directions would actually affect the subjects' descriptions. She discovered that they did. She worked out her statistical data, and wrote her paper, "Testing the Effect of Suggestions on Subliminal Perceptions," to enter in the Science Talent Search.

Amanda was chosen as a Science Talent Search semifinalist. Amanda's paper was selected to be presented at the regional Science and Engineering Symposium. Amanda discovered that many colleges were interested in her work. She received an academic scholarship to the college of her choice. In addition, she was allowed to begin advanced work in her major and was able to graduate in only three years.

Chapter 3
Getting Started

Getting started on the science research project that you plan to enter in the Science Talent Search or the International Science and Engineering Fair might seem a formidable task. In truth, it is rare to just wake up one morning and say, "I know the project that I want to do." Most students have no idea how to find a project topic, much less a very specific science research project.

An important first step is to start a scientific journal. The journal is simply a notebook that will become your resource book. Many people like to use a hardbound notebook to ensure that the pages do not get lost. Other people favor a lab notebook, which is a hardcover notebook with graph paper pages. The graph paper helps with your calculations and graphs of your lab data. You should keep a list of all the books and articles you have read, along with a short summary of each in your journal. In addition, your scientific notebook will be a diary of the daily and weekly progress of your project and your research. In this way,

Keep a journal that will become your resource book. Your scientific notebook will be a diary of the progress of your project and your research.

you will be able to refer to early readings and experiments in order to write and present your paper.

There are several things to keep in mind as you get started on your science research project:

1. Your project must be an experiment rather than a simple written report.

2. Keep your project as simple in design as possible. Research projects can range from the most elementary to the most complex. It is better, however, to think small rather than big. When you finish your small proposal, you can start a second experiment to either confirm or expand your results.

 Many students that have started with a project that is too large and too complicated discover that they never actually begin the experiment. They either become distracted by the process of understanding the technical nature of the proposal or encounter too many difficulties in setting up the experiment.

3. Always adhere to the scientific method. This means that you will state a hypothesis for your project. After that, you will design and conduct an experiment to reject or accept your hypothesis. (See Chapter 4 for detailed information about designing a research experiment using the scientific method.)

Your research project will be an assignment you do for yourself, not for anybody else. Completing a research project will take many hours of work in the laboratory. You will be working on your project, setting up the experiment, and collecting and analyzing data. If you do not like the project—indeed, enjoy it passionately—you will not complete it. If you are working in areas that are interesting to you, however, the endless hours will not seem like work but will be fun and rewarding.

The best way to get started on a science research project is to become involved in as many science activities as possible. It is important to become aware of all the opportunities that are around you. It is quite likely that after participating in many interesting experiences, finding the actual project you want to work on will just "seem to happen."

The following list of suggestions will help get you started in the world of science research:

1. Read as much as possible. Choose a broad topic, such as microbiology, chemistry, or engineering. Then go to the library or the Internet and research the subtopics that

comprise your main subject area. Look into professional journals, both in the library and online, and magazines as well as the books and textbooks that you will find. As you read, do not look for a specific project or experiment, but rather read to gain background information about your subject area and the research that has been conducted in the past. Also, become familiar with the scientists in your field.

2. Attend the scientific lectures, symposia, and conferences in your area.

3. Visit as many local industries, utility companies, and museums as possible. The computer, quality control, and design and engineering departments are interesting areas to visit.

4. Discover the student science programs in your area. There may be Saturday morning workshops, lab experiences, lectures, or even courses for high school students.

5. Visit the science departments at your closest university. Call and speak to professors in the areas that seem interesting to you. Many professors will be glad to discuss their own experiments with you. And, quite often, these professionals will invite you to visit their labs and even invite you to help work on an experiment.

6. Find out about the kinds of science courses offered at your local university. It might be possible for you to take a course there, either to learn about a new area or to enrich your knowledge about an existing topic.

7. Attend the science fairs that will be held in your town, county, and state. At these fairs other high school

students will present their own research projects. The presentations will give you many ideas for research experiments and poster presentations.

8. Find a mentor. Survey the people around that can help you with your project. In many high schools, science teachers have advanced degrees in specialized areas or have particular interests. Mentors can also be found at universities and in industry.

Science Clubs

You may discover that some of your friends also want to enter a science research competition. If that is the case, then you may wish to form a science club. A science club may become involved with all of the previous activities and many others as well. The following are some of the projects the club might wish to consider:

1. **Guest speakers.** You may wish to invite professors, scientists, researchers, or students to your meetings. The scientists and professors can discuss their work and careers. The student guest speakers could talk about their projects and their experiences at science research competitions.

2. **Brainstorming sessions.** The club members can pool their ideas for science research projects. They can share their projects with one another, discuss the successes and failures that they have encountered, solve problems together, and present their results to one another.

3. **Work on a major project together.** Each student would be responsible for one small aspect of the project so that at the end the group will have completed an

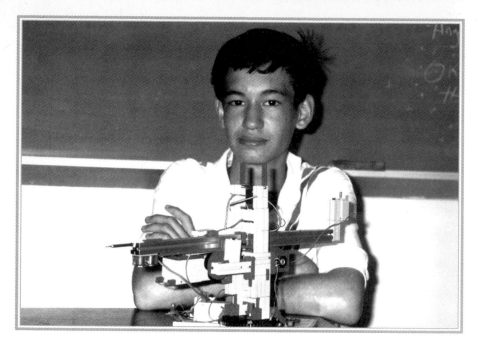

A student is shown with his successfully built robot for the Duracell Scholarship Competition.

extensive project beyond the scope of a single person. Several student clubs have successfully built telescopes and robots.

4. **Discuss current events in science.** For material to discuss, refer to the weekly articles in many national newspapers and science magazines. In addition, you can view films and videotapes on science topics.

5. **Host a competition.** This can be a small competition just for the members of the club. Or you could decide to host a major science fair for your entire county or region. If your school or community already hosts a science fair or other science competition, club members could help set up and host the fair.

6. **Student mentors.** Science club members can visit the elementary and junior high schools as well as classes in their own high school to give lectures and demonstrations. They can discuss their experiments and experiences while participating at research competitions. In addition, they can show other students how to make presentations, create good posters, and get started on their own science research experiments.

7. **Start a school science magazine, yearbook, or newsletter.** This student publication may contain student entries, articles of general scientific interest, winning science research project papers, puzzles, and other items of scientific interest.

Students in a science club look at weekly science articles in the library.

8. **Neighborhood project.** The club may want to work on a major science-related activity for the neighborhood or community. Some projects to consider might be cleaning up a lake, preserving wildlife, or establishing an organic garden.

9. **Community night of science.** This evening event can be sponsored by the club to present lectures by students, scientists, professors, and industrial leaders. These people can discuss their completed research projects, careers, and current projects. Students and scientists will be able to meet and exchange ideas about projects and areas of expertise.

Try to get involved in as many of these activities as possible. The more you do, the more you will see the kinds of research experiments in which scientists and high school students are involved. From these experiences, you will discover that you will find a specific area that seems most interesting to you that you can investigate for your own research project.

It is also important to know about the science research projects other high school students have completed in the past. Some winning titles from past Science Talent Search competitions and International Science and Engineering Fairs can be found in Appendix B.

Chapter 4

Doing the Research Project

Science research projects are very special projects. They follow the format that is used by research scientists in laboratories all over the world. Research experiments conform to the scientific method and entail these six steps:

Step 1: State a hypothesis to define the problem you want to solve.

Step 2: Design an experiment that will allow you to accept or reject the hypothesis.

Step 3: Build and assemble the experimental setup.

Step 4: Conduct your experiment and collect the data.

Step 5: Analyze the data using statistical methods.

Step 6: Write and publish the paper.

Following these steps will allow you to create a project that will be of the appropriate quality to enter in any of the prestigious science research competitions. The six steps are looked at in detail in the following pages.

State a Hypothesis to Define the Problem You Want to Solve

To begin, read! Find the background information related to your research area. Remember to consult current periodicals, journals, and the Internet, as well as books. Try to discover whether or not other people have attempted an experiment that is similar to the one you want to do and the kinds of results they have obtained.

Now you are ready to formulate your own hypothesis. The hypothesis will be a statement asserting the problem you will

A computer search can help you get started on your project.

be investigating with your experiment. As you work on your science research project, you will hope to either prove or disprove this statement.

Design an Experiment that Will Allow You to Accept or Reject the Hypothesis

Now that you have stated the problem, decide on the type of data that you will need to collect and how you will go about gathering it. Start by writing an outline of the procedures you plan to use. Plan a tentative schedule. Then think about the equipment and lab setup you will need. Determine whether special apparatus must be purchased or borrowed and whether a special arrangement must be constructed. If you will be working with live specimens, such as plants, determine where you will store them and how you will care for them.

Ascertain the rules of the competition you are entering for research involving vertebrate animals, human subjects, recombinant DNA, and tissue. Fill out the appropriate forms for these areas. Discover the safety measures that you must take with your experiment. Find out whether you will need consent forms for projects involving human subjects. For projects in the social science area, develop your questionnaire and define the subject pool you wish to test.

Remember to design your experiment so that only one variable at a time is being changed or tested. In this way, you will know how each variable affects the results. If too many factors are changing at the same time, you will see only the result, but not the reason.

Build and Assemble the Experimental Setup

Assemble the equipment for your experiment. You will want to hook up and test your apparatus and other hardware so that you will be able to conduct your experiment without frequent breakdowns. While conducting your experiment, you may encounter any number of problems with the equipment or lab setup, so modifications will have to be made to your original plan. For projects in the social science area, create or locate the survey form that you will be using for your experiment.

Conduct Your Experiment and Collect the Data

Carefully follow the procedures you have previously defined. As you conduct your experiment, you will receive feedback on your experimental design. You may find that you must revise your procedure or redesign parts of your lab setup because of equipment failure and breakdown.

If you are working on a social science research project that involves a questionnaire, this is the time to give out the questionnaire and collect the responses.

Analyze the Data Using Statistical Methods

Interpret and analyze the data you have collected from your experiment. First, describe your data, using tables, charts, and graphs. Then, interpret the data using statistical methods. In addition to the mean and standard deviation, apply the appropriate statistical test, such as t-test or chi-square test. This

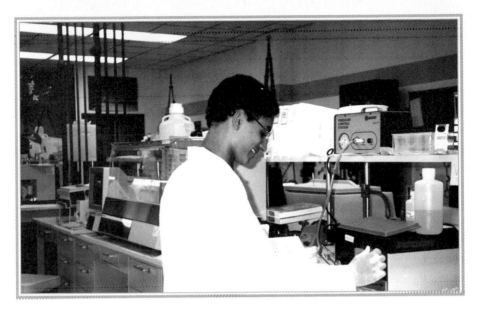

Observe your experimental trials closely and carefully record all your data.

will enable you to make a scientific judgment about whether you have proved or disproved your original hypothesis.

Your statistical analysis can be accomplished by using a calculator and formulas or by using one of the commercial statistical computer packages.

Write and Publish the Paper

Your work is not complete until you write up your results to share with other scientists and researchers. See Chapter 5, "Writing the Research Paper and Abstract," for complete details.

The following accounts show you how two students, Carl and Amanda, used the scientific method as they worked on their independent research projects. As you follow each of these papers, note that although the students were working in different areas and on different kinds of projects, they both used the same method. (The students who wrote these papers are described in Chapter 2.)

Carl's Paper: "Control of Growth Patterns: The Response of Bean Roots to Centrifugal Force and Gravity"

Before beginning his project about plants and their responses to gravity and centrifugal force, Carl spent time reading periodicals, journals, and books. In addition, he discussed his forthcoming project with several scientists at a local federal research laboratory. Here is a copy of the abstract to give you some background information about Carl's paper:

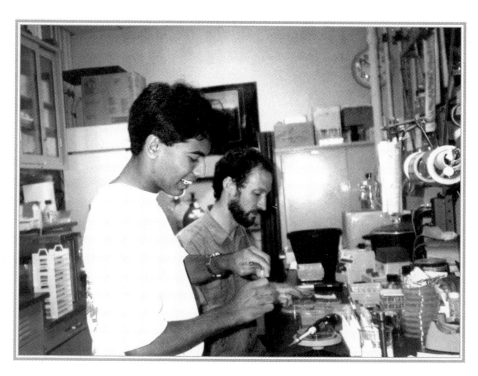

Students work very closely with mentors on their projects.

A plant placed horizontally responds to gravity and reorients itself to grow vertically. I wondered whether plants respond to centrifugal force. I built a centrifuge from a bicycle wheel and an electric motor. Bean seedlings growing in soil were centrifuged. Seedling roots did in fact respond to centrifugal force even when their response to gravity was blocked. Thus, the response of beans to centrifugal force differs in some ways from their response to gravity.

Carl's hypothesis stated, "I wondered whether plants respond to centrifugal force."

Carl broke his planning down into three parts: building the centrifuge, working with the plants, and performing the experiment.

The Centrifuge

The centrifuge was to be created from a bicycle wheel mounted on a wooden frame that would be driven by an electric motor.

Plants and Soil Mix

Tendergreen or Tendercrop (bush type) bean seedlings were chosen for the experiment. The bean seedlings were to be germinated at room temperature on damp paper towels and then transferred to a soil mix.

The soil mix was to consist of two parts commercial potting soil to one part vermiculite.

After the roots were about one to two centimeters long, they were to be transplanted into plastic freezer cartons.

Performing the Experiment

1. A control group of plants would be grown.

2. The cartons would be placed on the centrifuge approximately nine inches from the center of the wheel.

3. The RCF (g force) would be calculated according to the formula: $RCF = KRF^2$, where $K = 2.838 \times 10^{-5}$, R = radius in inches, and F = frequency in revolutions per minute.

4. Four to eight seedlings would be centrifuged in each experiment.

5. After centrifugation, the seedlings would be removed from the soil and inspected. Carl's reading had indicated that the roots would remain rigid and hold their shape when removed from the soil.

6. The roots' orientation would be noted.

7. The seedlings would be rinsed and photographed.

8. The roots would be preserved in Carnoy's solution (three parts ethanol and one part acetic acid).

The centrifuge was built according to the plans. It was discovered that the centrifuge wheel wobbled excessively at high speeds. The problem was solved by reinforcing the frame and creating new supports for the motor brackets.

It was discovered that the desired g force, usually 1 x g or 4 x g, was dependent on the speed of the centrifuge. The number of revolutions required could not be counted by eye. To determine the speed of the centrifuge, a piece of cardboard was taped to the wheel; it stuck out slightly so that it clicked against the motor's mount each time the wheel revolved. Thus, with the aid of a timer and a metronome, the proper number of revolutions could be set to yield the desired g force.

The tests began and several more problems were encountered:

1. The soil mix turned out to be very sensitive to centrifugal force. This was detected when the soil was flung from the spinning containers so that it covered everything within a wide range. The sandwich wrap covers were replaced with snap-on lids.

2. In dry soil, the beans spun out of position. Therefore, before centrifugation, the soil was dampened and prerun.

3. The plan to preserve the harvested seedlings in Carnoy's solution had to be abandoned because the solution distorted the roots.

Like Carl, Chia tested her project to work out problems.

Once the experiment began to work properly, the roots were removed, the orientation was noted, and the seedlings were rinsed and photographed.

The amount of change from the vertical for bean roots was noted. Correlation coefficients between change in direction and increased g force were determined. Change in growth direction increased with increasing force. Carl wrote his paper and submitted it to the various research competitions.

Amanda's Paper: "Testing the Effect of Suggestions on Subliminal Perceptions"

Before Amanda began to work on her project, she spent time searching the Internet and reading periodicals, journals, and books about subliminal suggestions. In addition, she enrolled in a first-year psychology course at a local university. Amanda then spent the summer after her junior year studying psychology at a university summer program for high school students. For some background information about Amanda's paper, here is a copy of the abstract:

In this psychological investigation, subjects were tested in an attempt to find out whether or not a blatant suggestion as to a "correct answer" would actually affect the subjects' responses. Subliminal perceptions are sense perceptions, in this case, visual, which are weak, but nevertheless register somewhere in a sensory form. My goal was to see if I could influence a subject's perception by offering them a sense of what they "might" see.

Statistical analysis was used to prove that there was a significant difference between the data for the control group and that of the experimental group. The Mann-Whitney U Test showed that the hypothesis, that a suggestion would affect the subjects' responses, was valid.

Amanda's hypothesis stated: "I wondered whether or not a blatant suggestion as to a 'correct answer' would affect a subject's responses."

Amanda designed the following experimental protocol:

1. Choose the subject pool. The chosen group should be of high intellect with all students members of an honors science course in Amanda's high school.

2. Obtain a copy of Rorschach Test to test subjects.

3. Obtain index cards to record data.

4. Decide on the statistical methods to be used to analyze the data.

Amanda discovered that she was not able to obtain copies of the Rorschach Test to use for the project because of restrictions placed on the usage of the actual Rorschach plates. Therefore, after consulting with psychologists at a local university, Amanda designed her own set of ink blot plates.

The ink blot plates were evaluated by an institutional review board made up of high school and college teachers. The subject pool of honors students was selected and notified. Informed consent statements were provided for each of the subjects.

Performing the Experiment

1. Each subject was shown a series of ten ink blots, very similar in design to those of the Rorschach Test.

2. The test was administered to the control group in a one-on-one situation. Responses were recorded individually on three-by-five-inch index cards.

3. The same series of plates were then presented to an

experimental group. However, there were two changes in procedure. The plates were photocopied, and each subject was presented with his or her own package. Responses were then recorded individually by the subjects in the space just below the ink blots.

4. A different set of directions was read to the control and experimental groups. The control group was asked to describe the ink plates. The experimental group was given the same set of directions with the following sentence added: "This test will attempt to determine how violence is perceived."

Tables of the collected data were then created with the headings: "Results from the Control Group" and "Results from the Experimental Group." The standard deviation for the control and experimental groups was calculated.

The data was submitted to a statistical analysis by the Mann-Whitney U Test, which was selected because of the nature of the collected data.

With the statistical analysis complete, Amanda was able to accept her original hypothesis: There was a significant difference between the experimental and control groups. A blatant suggestion affected the subjects' responses. Amanda then wrote her paper and submitted it to the various research competitions.

Chapter 5
Writing the Research Paper and Abstract

The goal of a research scientist is to publish his or her work. A researcher must be able to communicate the results of experiments to fellow scientists. In this way, the successes and failures the researcher has achieved are shared with the entire scientific community. Other scientists are able to see the methods the researcher used and the ideas tried, and know where to begin their next set of experiments.

Therefore, the real work begins after you have completed your project and tabulated the results. No matter how spectacular your experiment and its results, the project is not complete until the descriptive paper has been written. The paper must clearly and concisely describe the results you have obtained and the meaning of those results.

After you record your experimental results in the form of a paper, you would like others to read and share your results. The way to do this is in a scientific journal or, in the case of a high school student, a competition such as the Science Talent Search. There are scientific journals in every field of science. A journal is created by a peer review system. Your paper will be sent to one of the reviewers who is a scientist in your field of study. He or she will read the paper and recommend whether to accept your paper for publication.

Your research paper for publication or presentation should follow very specific rules and have a very definite format. Keep the following points in mind:

1. Do not write too much when you write your paper. For instance, the Science Talent Search requires that your paper be from five to twenty pages in length. This means that the judges are looking for quality rather than quantity. Make sure that everything you include is relevant to your actual experiment. Do not fill your paper with random and aimless thoughts. Always write concisely and to the point.

   If you feel that you must add an appendix to explain your methods or certain procedures, then do so. But keep in mind that your report should be able to stand by itself.

2. Make sure that your final report is printed, double-spaced on white paper. All the drawings must be done neatly in black. Simply staple your paper once in the upper left-hand corner. Do not include fancy covers.

3. Always use correct grammar and punctuation.

4. The abstract should be included in your paper immediately after the title page.

5. Research papers are usually arranged with the following section headings.

Title

Abstract

Introduction

Materials and Methods

Results

Discussion

Conclusions

Literature Cited

Acknowledgments

6. Do not begin each section on a new page. Each section should follow immediately after the previous section.

7. Do not use a table of contents.

Each section of the paper is described on the following pages.

Title

The title must be self-explanatory. That is, anyone reading it should be able to clearly understand the purpose of your experiment. For example, a title such as "A Biology Lab Report" tells you nothing. The reader has no idea whether the research involves an animal or plant or what is measured or tested. An example of a good self-explanatory title would be: "The Effects of Light and Temperature on the Growth of *Artemia salina.*"

Note that the title specifically states three things:

Your mentor can help you polish your final research report.

1. The environmental factors that were manipulated—in this case, light and temperature.

2. The parameter of the organism that was measured—in this case, growth.

3. The specific organism that was used—in this case, *Artemia salina.*

Introduction

The introduction clearly states your hypothesis so that the reader can understand the question or questions that you are trying to answer.

In order to accomplish this, you must provide the reader with the background information or history of the problem. This is usually accomplished by a literature review. Whenever possible, primary sources should be used. The primary source is the original paper in which the important information was initially published.

It is virtually impossible to start your own research with primary sources unless you have been working in that particular scientific field for a long time. So the best way for you to start would be to go to the Internet or a general sourcebook such as an encyclopedia or textbook. Read everything in these books that is pertinent to your experiment. Then check the footnotes and, "Works Cited" sections in these books as well as your regular library resources (such as the periodical guide) and the Internet for further information about your topic.

In this way, you will be able to backtrack from very general information about your topic to the very specific. You will discover as you go along that you will be acquiring information so that what originally looked like an impossible amount of information to absorb because of its great detail and complexity is becoming quite understandable. When you finally backtrack to the author of the original works that have been cited on the Internet and in many texts and encyclopedias, you have arrived at the primary source.

Another way to track down background information is to do an author search either on the Internet or at the library. In this type of search, you begin with the author, rather than the topic. In your initial reference list, find those scientists who have made significant contributions to the area in which you are interested.

You can use key words to locate articles as well. Use your subject area or key word in a computer search at the library. This technique will help you find books and articles on the topics you are considering for your research.

Dr. Norman Goodman assists a student with her literature review.

In the introduction, you may wish to include how your present experiment will help expand the knowledge of your general area of study.

Materials and Methods

The materials and methods section is a detailed description of how you did your work. Your description must be exact and complete so that anyone who wants to do so can repeat the experiment. Therefore, this section will include the experimental design, the apparatus, the methods of gathering data, and the control that was used. In addition, if you collected any specimens for the study, you must say where and when the material was collected.

A photograph or a diagram can be used as an aid to describe the apparatus or laboratory setup. This will often save a great many words and create a much clearer picture for your readers.

It is important to keep in mind the following rules:

1. Do not list your equipment. This would make your paper sound more like a lab report than a scientific paper. Your methods should be described in full sentences and paragraphs.

2. Do not describe standard methods in detail. For example, if you used the statistical t-test, just state that the method was used to test for significance. The details of the method can be placed in an appendix or listed as a reference.

3. The materials and methods section should be written in the past tense. The scientific paper is a report on something that the researcher has already completed. The paper should read like a narrative in which the researcher "tells his story."

4. Discuss weaknesses and limitations of your method.

Results

The results section describes what happened after you completed your experiment. This is a very straightforward section. The results are simply stated for the readers to observe. Do not include any interpretations, conclusions, or value judgments. If possible, the results should be presented in a graph or table.

It would be a good idea for you to learn how to use a computer program to analyze your data for the results section. The program will enable you to perform many types of statistical analysis and data management tasks. For instance,

Computers are useful for the statistical analysis of data.

you will be able to use simple computer commands to type in your data and then create graphs and tables.

The tables and graphs must be accompanied by narrative text. Each graph or table must contain a self-explanatory title as well as enough information to describe the results being presented. If tables and graphs are correctly labeled and titled, they should be able to stand on their own if read apart from the paper.

If you have worked on many different experiments, you might want to break up your results section into several parts.

Discussion

The discussion section will explain what the results mean. You will describe any patterns that emerged, any relationships that appeared to be meaningful, and any correlations that can be discerned. This can often be accompanied by a statistical analysis of the data. A statistical analysis can be completed by working with a statistical computer package or statistical calculators.

The discussion section will describe the reasoning and logic that led you to your conclusion.

You must include any explanation about why you think the results turned out different (if they did so) from the way you expected (e.g., the timer broke down, you forgot to add a chemical). In addition, the discussion should include any reasons that the results were either different from or similar to any related experiments by others. This is the time to discuss the complications that arose during your experiment (e.g., the cells stopped growing, the silver stain would not take).

In the discussion section, it is important to compare your results and conclusions with the information already known about the problem, such as past experiments or observations conducted by you or by others. In summary, you are drawing conclusions about the meaning of your data and are explaining why you reached those conclusions.

Conclusions

The conclusions section is simply a summary of the conclusions and important discoveries that you have made while working on your research. In a sense, this section is a recapitulation of the discussion and results sections except that this time you are just making a list of your important conclusions and discoveries without giving the reasons that you have reached these conclusions.

This section is the essence of all that you have discovered with your research. Anyone should be able to read the introduction and conclusions sections and quickly determine exactly what has been accomplished without fully knowing the details of how the research was conducted.

The conclusions section includes any further research that might be conducted in your general area of experimentation.

Literature Cited

In the literature cited section, you will list all the references that you have cited in your paper.

All the background information that you included in your paper must be properly referenced. Numbered superscript footnotes are not usually used when citing sources in a science research paper. Instead, the author's last name and the date of publication are placed in parentheses in the text after the cited reference. If you have included the author's name in the text, place only the date of publication in parentheses.

However, footnote styles and rules can be quite idiosyncratic in different scientific journals and competitions. So before submitting your paper for publication in a journal or in a competition, check the rules or the publication's style guide for citing sources.

The references are usually listed in alphabetical order by author at the very end of your paper. This listing will enable the people who have read your paper to make a general literature search on the general problem you have worked on. Make sure that you include only those references that were actually mentioned or cited in your paper. Any other background information that you read in learning about your experiment but did not cite in the paper is not included in this section.

Acknowledgments

In the acknowledgments section, you will thank all the people who helped you with your project. This section should be only one paragraph long and placed on its own page.

Writing the Abstract

The abstract is a one- to two-paragraph summary of your entire paper. This summary is always written on a separate sheet of paper and placed after the title page and before the main body of the paper. It should consist of no more than two hundred words. The abstract includes your hypothesis, a summary of your problem, and the results that you obtained. The abstract is written after you have completed your project and written your paper.

Contest judges will receive their initial impressions of your complete paper from the abstract. In fact, many times, the decision whether to read the paper is made after reading the abstract. If the abstract does not describe the project, list the hypothesis, or discuss the results of the project, the judge will probably not receive a good first impression of your paper.

Often, after you have finished an experiment and written a paper, you may wish to try to have the paper published. However, rather than reading an entire paper submitted to a journal for publication, the peer review panel first reads the abstract of the submitted paper. This abstract gives the judges a summary of the problem to be solved, the hypothesis that you worked on, and the results that you obtained.

Following are some examples of winning abstracts from the Science Talent Search and International Science and Engineering Fair competitions:

"A Methodology for Projecting the Population Center for the United States"

The center of population is that point at which an imaginary flat, weightless and rigid map of the United States would balance if weights of identical value were placed on it so that each weight represented the location of one person on the data of the census.

Through intensive research, I was able to create a computer program that, with the data that I compiled, projected the next center of population: 37° 44′ north latitude, 91° 4′ west longitude, 10 miles due east of Viburnum, Missouri. This marks the largest single southward leap of the population center.

A student completes her research for the Science Talent Search competition.

"Movement of Neutrophils Stimulated by Chemoattractant"

Neutrophils are cells that migrate from the blood into infected tissues and destroy bacteria. They are attracted to the infection by bacterial products that bind to the neutrophil's surface. In these experiments, conditions of temperature, acidity and salt concentrations, and types of tissue molecular constituents that enhance movement of stimulated cells were identified. Knowledge of these effects, shown in quantitative terms using a specially designed cell culture apparatus, may permit development of methods to enhance neutrophil response to infection.

For some competitions, such as the Science Talent Search, you must also include a seventy-five-word summary written in language that can be understood by nonscientists. This summary should follow the title page.

Chapter 6

Presenting the Paper

After you have completed your experiment and written your paper, you may be invited to make a presentation of your research. The presentation is a very important aspect of the total project. It is not enough to conduct an experiment and write a good scientific paper; you must also be able to communicate your results with others. You must be able to extract the most important facts and figures and present these ideas and findings so that many people can understand the significance of your results and conclusions.

Most commonly, presentations are made to a large audience by using slides or view graphs, or to a small group by using posters.

Speaking Before a Large Audience

When you are addressing a large audience, you must set your talk to the audience level. The presentation must be carefully planned and outlined. Keep in mind that you should break it into three simple parts:

1. Tell the audience in an introduction exactly what you plan to tell them in the next few minutes. This should be a summary of your abstract.

2. Inform the audience of your important results or conclusions. Be clear and concise. Do not get carried away with too many details of how you conducted your experiment. Do not include extremely technical jargon unless your audience consists totally of experts on your topic. Do talk only about the pertinent material. A good section of your paper to paraphrase for this section of your talk is the conclusions section.

3. Tell the audience what you have just explained to them. Simply summarize the results and conclusions of your research.

While talking to a large audience always talk slowly and carefully.

It is a good idea to use view graphs (overhead projections), a computer presentation such as Microsoft® Power Point, or slides to help explain your talk. Each slide or view graph should contain only a single idea or theme. The view graph must be self-explanatory. That is, the people in the audience should be able to look at it and understand the sentence or concept before you begin to talk. You must clearly label any tables or graphs that you present as slides or view graphs.

Many times, the slides or view graphs constitute the outline for the talk. You should be able to move from slide to slide without hesitation. The use of view graphs or slides should free you of the need to use an outline or note cards.

When writing your talk, keep the level of the audience in mind. Have a friend or mentor help you with your presentation.

When using slides, make sure you know how to use the remote control. Practice with your projector before you give your talk.

When you are using view graphs, do not stand next to the projector. Instead, have another person place the view graphs on the projector. Then, stand near the screen and use a pointer to indicate the important aspects of each view graph.

Try not to plan your talk so that you will be writing and creating the view graphs as you talk. This can be quite distracting to your audience. At the same time, because of nervousness, you may find it more difficult to write what you had planned or may even forget exactly what you intended to mention.

Remember to look at your audience and not just at the screen or pointer. And best of all, try to smile as you talk.

Smiling forces you to look at your audience and has the effect of calming your nerves. Remember to be yourself: You did the project. You wrote the paper. You are in charge. Relax and enjoy what you are doing. Everyone wants to hear what you have to say. If you make a mistake while talking, just stop and start again.

A One-on-One Poster Presentation

Quite often a poster presentation involves you and a single judge. Usually you will have between ten and fifteen minutes to present your work. It is important that you use this time wisely. The first two minutes are the most important. In that time, you must impress the judge with the quality of your work and the conclusions you have reached.

Your presentation must be carefully planned and outlined. Do not think that a good presentation can simply be spontaneous. An excellent presentation will appear to be spontaneous, but, in fact, it will be precisely organized and coordinated. It is important to practice your talk with your peers and teachers.

Break your talk into three parts as you did for the audience presentation. Remember to look directly at your judge and not just at your poster display. Make sure that you maintain eye contact and try to smile as you talk. If you make a mistake, stop and start again.

You will have to set up a display with posters showing photographs, graphs, tables, and figures relating to your project. When you make your display, keep the following ideas in mind:

During a one-on-one presentation look directly at your judge and not just at your poster display.

1. The display should make an immediate impact on your judge. It should be set up so that with a quick glance your judge will be able to assess the merit of your project.

2. The posters must be self-explanatory. That is, the judges should be able to look at your display and understand much of your project before you begin to talk.

3. You can use your posters as an outline for your talk. You should be able to move from poster to poster without hesitation and without an outline or note cards. In your talk, make sure you refer to all the material that you display.

4. Your display should be as attractive as possible. However, attractive layout and bright colors are not enough; you must include substance and quality.

5. Any item included on your posters should be simple and clear. Each picture or table should contain only a single idea or theme. You must clearly label any tables or graphs that you include.

6. Include in your display a copy of your research paper, your journal notebooks, computer programs, and any other relevant written materials.

For the International Science and Engineering Fair (ISEF) and other poster presentation competitions, make sure you follow the guidelines and observe the restrictions of the contest. Each competition is quite specific about the size of the display. For example, in the ISEF, your exhibit size is limited to 76 cm (30 in) deep, front to back; 122 cm (48 in) wide, side to side; and 274 cm (108 in) high, floor to top. You will be given a table 76 cm (30 in) high. Any exhibits exceeding these dimensions will be disqualified at the ISEF.

Answering Questions

Just as important as the formal presentation of your paper is the question-and-answer period that you will have at the end of your talk or poster presentation. At this time, your understanding of your results, your experimentation, and even your errors and mistakes will come under fire. You must be able to answer questions about your research and back up in words the conclusions that you have reached. You must be prepared for any and all kinds of questions. A way to do this is to actually anticipate the kinds of questions that you will receive.

If you are giving a talk before an audience or during a competition, and no one asks questions, be prepared to step in yourself with such sentences as the following:

- Although there are no questions, perhaps you would like to know why I chose to use this method, this medium, this kind of experiment, etc.

- You may want to know why I think these results are inconclusive, unusual, excellent, etc.

- I was surprised to discover . . .

It is most important to always be honest when answering questions. Therefore, if you honestly do not know the answer or never thought to look at the problem in that particular manner, do not be afraid to say, "I do not know." In fact, sometimes, the judge may even ask you a trick question to see whether you actually know exactly what you are talking about. It may be a question that only a Ph.D. or graduate student

Answer the judges' questions honestly. Do not be afraid to say, "I don't know" when you do not know the answer.

may understand. Or the judge may know that there is no answer to a particular question.

Do not be afraid to take a little time when answering a question. Do not start talking without taking a moment or two to think out your response.

A terrible response to a question you find difficult would be to start your talk over again. Avoid this at all costs. If you have already stated the answer in your talk, do not repeat the exact words. Find synonyms for your original ideas. Sometimes, the judge or audience is just looking for clarification of your ideas.

While answering questions, do not lose your cool. Do not become impatient or intolerant of the person asking them. Try to give the impression that you are always in control. After all, you are the person giving the talk. Everyone has come to listen to you and find out the results and conclusions of your experiment.

Chapter 7

Judging Guidelines

You have finally completed your work. After all the hours you have spent doing your research, working in the library, making endless phone calls, writing letters to all kinds of mentors, and then finally writing the paper, you have set up your project at a science fair and you are about to be judged.

At many science and research fairs, the judges will come and talk directly to you. Quite often there are several rounds of judging. At regional and local science fairs, you may meet only one or two judges, but at the final round of the national fairs, such as the International Science and Engineering Fair, you will meet as many as seven!

Before you speak to your judges, prepare! It is a good idea to have a ten- to fifteen-minute talk prepared. However, before you begin your discussion with the judges, know the criteria that will be used to award points in different categories in the competition that you have entered.

Many competitions publish a guide explaining how points will be allocated during a competition. Study the guide to find out the kinds of questions that will be asked, the type of display the judges will be looking for, and the type of paper and abstract that will be most successful.

Following are the judging guidelines for the International Science and Engineering Fair and the Science Talent Search. Although competitions usually publish their own set of criteria and procedures for the judges, the guidelines for the ISEF are among the most comprehensive. As you prepare to enter a competition, check yourself against the ISEF judging criteria. This should prepare you for any interview in any competition.

The Science Talent Search

The major goal of the Science Talent Search is to find and encourage students with a creative talent for science and engineering.

The selection process for choosing the three hundred students in the Honors Group and the forty winners is as follows:

The judges confirm that the student's application is complete (i.e., that the student has included a science project report, a completed entry form, a high school transcript, and a list of standardized test scores).

Each complete application is reviewed by at least two independent evaluators. Their task is to

1. Make an independent evaluation of the science project report.

2. Evaluate the entry form for information that reflects the student's promise as a creative scientist.

Grace Lau and Christopher Gerson, both of Setauket, New York, were finalists in the Science Talent Search. Grace's project was on the effects of green tea, and Christopher's project studied earthquake motion.

3. Review the student's relative standing in his or her high school graduating class.

4. Review the student's scores earned on standardized tests.

After the three hundred students in the Honors Group are selected, forty of these students are chosen as winners. The forty winners then go to Washington, D.C., to face judges and evaluators who choose the top ten winners.

International Science and Engineering Fair

The International Science and Engineering Fair awards points in the following manner. The same scoring procedure is used at the regional and national levels of the competition. Students may enter either as individuals or as members of a team project. Teams may have up to three members.

	Individual	Team
Creative Ability	30 points	25 points
Scientific Thought/Engineering Goals	30 points	25 points
Thoroughness	15 points	12 points
Skill	15 points	12 points
Clarity	10 points	10 points
Teamwork	——	16 points

Creative Ability

In awarding the points for creative ability, the judges consider whether the project exhibits creativity and originality in each of the following ways:

1. The original question or hypothesis

2. The approach to solving the problem

3. The analysis of the data

4. The interpretation of the data

5. The use of equipment

6. The construction or design of new equipment

Scientific Thought/Engineering Goals

In awarding the points for scientific thought and engineering goals, the judge seeks answers to the following questions:

For Scientific Thought

1. Is the problem stated clearly?

2. Was the problem sufficiently limited so that it was possible to attack it?

3. Was there a procedural plan?

4. Are the variables clearly recognized and defined?

5. If controls were necessary, was there a recognition of their need, and were they correctly used?

6. Are there adequate data to support the conclusion?

7. Are the limitations of the data recognized?

8. Does the student understand how the project ties in with related research?

9. Does the student have any idea of what further research is indicated?

10. Did the student cite scientific literature, or cite only popular literature?

For Engineering Goals

1. Does the project have a clear objective?

2. Does this objective have relevance to the needs of the potential user?

3. Is the solution

- workable?

- acceptable to the potential user?

- economically feasible?

4. Can the solution be successfully utilized in design or construction of some end product?

5. Does the solution represent a significant improvement over previous alternatives?

6. Has the solution been tested to see whether it will perform under the conditions of use?

Thoroughness

In order to award the points for thoroughness, the judges will be looking for answers to the following questions:

1. Does the project carry out its purpose to completion within the scope of the original aims?

2. How completely has the problem been covered in the project?

3. Are the conclusions based on a single experiment or on replication?

4. If it is the kind of project where notes were appropriate, how complete are they?

5. Is the student aware of other approaches or theories concerning the project?

6. How much time was spent on the project?

7. Is the student familiar with the scientific literature in the field in which he or she is working?

Skill

In order to award the points for skill, the judges answer the following questions:

1. Does the student have the skills required to do all the work necessary to obtain the data that support the project?

2. Where was the project done? What assistance was received from parents, teachers, scientists, or engineers?

3. Was the project carried out under the supervision of an adult, or did the student work largely on his or her own?

4. Where did the equipment come from? Was it built independently by the student? Was it part of a laboratory in which the student worked?

Clarity

In order to award the points for clarity, the judges answer the following questions:

1. How clearly is the student able to discuss the project?

2. Has the written material been expressed well by the student?

3. Are the important phases of the project presented in an orderly manner?

4. How clearly are the data presented?

5. How clearly are the results presented?

6. How well does the project display explain itself?

7. Is the presentation done in a forthright manner, without cute tricks or gadgets?

8. Was all the work done by the student, or was assistance received from his or her art class or others?

After working for many hours on their research project, students will have the laboratory skills that impress the judges.

Teamwork (Team Projects Only)

1. Are the tasks and contributions of each team member clearly outlined?

2. Was each member fully involved with the project, and is each member familiar with all aspects?

3. Does the final work reflect the coordinated efforts of all team members?

Judging Criteria Used at Many Local Science Competitions

Many other competitions have other judging criteria. For example, the Long Island, New York, Science Congress utilizes the following criteria for judging research experiments:

1. Quality of abstract	0–5 points
2. Originality of project	1–10 points
3. Formulation of hypothesis	1–10 points
4. Evidence of literature search	1–10 points
5. Experimental design	1–10 points
6. Implementation of design	1–10 points
7. Effectiveness of display	1–10 points
8. Oral presentation	1–10 points
9. Validity of conclusions	1–10 points
Total maximum possible	85 points

As you can see by comparing the judging at the Long Island Science Congress and the International Science and Engineering Fair (see page 68), many of the items judged are similar.

From the beginning of your research project, keep the judging criteria in mind. It is a good way to avoid disappointments when it is time to present your research. You will be prepared for all the project elements for which your work will be judged.

Chapter 8

To Enter Is to Win

We will finish with the story of Ben and Paul. Ben and Paul decided to join a research class.

They very quickly became involved in all kinds of projects.

First, they learned how to build balsa wood bridges. As they worked on their projects, they encountered many problems. Their first set of bridges was totally disqualified. But they continued. Paul entered the math fair. Ben entered the science fair. They joined a team to build a balsa structure for Odyssey of the Mind, a national problem solving competition. Together, they wrote an energy conservation proposal to create a more efficient windmill. Finally, in their junior year they each began a major research project to enter in the Science Talent Search. They combined their research projects with the ideas they used for their original windmill energy project.

In a research class, you work with a partner on a project.

Now they went at it with a vengeance. Together they did a lot of research on windmills. They each built a 1/50 working scale model of a Dutch windmill. They each read about and experimented with different blade diameters, increasing and decreasing the width of the blade from the base to the tip, changing the number of blades, adding and deleting a tail vane, and working with different wind speeds. They tested their windmills with the many different parameters and created reams and reams of data. Then the data was statistically analyzed. Finally, each wrote and presented a paper, and entered it in the Science Talent Search.

Did they win the competition? No, they did not win a major competition. But, in the bigger picture they have won it all. They learned a lot about windmills, alternative energy, building models, making presentations, and writing papers. They have become young men who are quite sure of their abilities in the science world. When it came time to write their college applications, they had credentials, awards and certificates, and great experiences! They looked forward to their college interviews.

Most of all, they knew that they had the ability to start, complete, and present a major research project. So, does it really matter if they never won a national contest? No, because just by working on a project they have already won. And so can you!

Appendix A

Major Science Competitions

Entering science competitions allows you to share the information you have gained from your projects and experiments with other students, teachers, and professional scientists.

The following is a list of the major competitions found throughout the United States, with a preface about local competitions.

Local Competitions

Before entering a major competition, it is a good idea to either enter or attend a local competition, in order to begin to learn about the process of competing.

Perhaps there is a science competition in your school. If not, talk to the science teachers in your school district to find out about the competitions that are available to you. If no science fairs exist in your area, your teachers will be able to locate the statewide competitions available to you and the requirements to enter.

Even if you cannot enter with your own project, do not skip the fair. Visit and see the types of projects other students in your age group and area are working on.

National Competitions

Competitions often change. Internet addresses have been included where possible to help you keep up to date.

The Dupont Challenge/ Science Essay Awards Program

Participants: Students in grades seven through twelve.

Format: Students are required to submit essays of six hundred to one thousand words on a scientific topic and its impact on society. Research is completed via a library or Internet search.

Awards: Educational grants.

For Information Write To:

> **Science Essay Awards Program**
> c/o General Learning Corporation
> 60 Revere Drive, Suite 200
> Northbrook, IL 60062–1563

Web Site: http://www.glcomm.com/dupont

Duracell Scholarship Competition

Participants: Students in grades seven through twelve. Students may enter individually or in teams of two.

Format:

Round I. Students design and build a device that runs on Duracell batteries. Students must send in a photograph, wiring diagram, and a two-page description about the device.

Round II. The top one hundred finalists (or pairs) are notified to send in their actual devices for the final judging.

Awards: Scholarships.

For Information Write To:

> **Duracell Scholarship Competition**
> National Science Teachers Association
> 1742 Connecticut Avenue, NW
> Washington, DC 20009

Web Site: http://www.nsta.org/programs/duracell.shtml

International Model Bridge Building Contest

Participants: Students in grades nine through twelve.

Format:

Round I. Regional. Students compete by building bridges of bass or balsa wood given certain specifications and constraints. The specifications change from year to year. Bridges are tested for efficiency. The builders of the most efficient bridges are the winners.

Round II. International. The two top winners from the regional event are eligible to participate in the national competition. The national competition is held in various cities around the nation.

Awards: Scholarships and prizes.

For Information Write To:

> **Earl Zwicker**
> Department of Physics
> Illinois Institute of Technology
> Chicago, IL 60616

Web Site: http://www.iit.edu/~hsbridge

International Science and Engineering Fair (ISEF)

Participants: Students in grades nine through twelve. Students may enter either individually or in teams of up to three students.

Format:

Round I. Regional. Students submit completed science, engineering, and math independent research papers to regional coordinators. Selected papers are then presented to judges. Two students and one team are named finalists from each of the regional competitions to compete in the international event. Some of the project categories are behavioral and social science, biochemistry, botany, chemistry, computer science, Earth and space sciences, engineering, environmental sciences, mathematics, medicine and health, microbiology, physics, and zoology.

Round II. International. At the INTEL ISEF, students set up poster presentations and are interviewed by a panel of judges. The international fair is held annually in different cities throughout the world.

Awards: More than five hundred fifty competitive awards are presented to the students.

Sponsor: The ISEF is administered by Science Service and funded by the INTEL Corporation.

For Information Write To:

Science Service
1719 N Street NW
Washington, DC 20036

Web Site: http://www.sciserv.org

International Science Olympiads

Participants: Students in grades nine through twelve.

Format: The International Science Olympiads are highly selective competitions in the fields of mathematics, physics, chemistry, biology, computer science, and astronomy.

Round I. A written test in any one of the given areas. This round is open to all United States high school students.

Round II. A select group from Round I is given a second written exam.

Round III. The top students are then selected to attend a national camp in their field of study for two weeks during the summer. The top four students from the camp are then chosen to participate in the International Olympiad.

Awards: The top students receive an expense-paid two-week trip to a study camp.

Web Site: http://www.olympiads.win.tue.nl

Junior Engineering and Technical Society (JETS)

1. Tests of Engineering Aptitude, Math, and Science (TEAMS)

Participants: Grades nine through twelve. Schools may enter teams of four to eight students on both varsity teams and junior varsity teams.

Format: The TEAMS exam is given in two parts. The first part consists of a series of objective (multiple choice) questions related to various engineering situations. The second part requires students to describe and defend their solutions to open-ended, subjective questions related to some of the problems they worked through in part one.

Awards: Vary from year to year.

2. Engineering Design Challenge

Participants: Grades nine through twelve.

Format: This competition requires teams of students to design and build a working model of a solution to a societal problem that is given to them. Students are given a set of specifications and constraints. The actual problem changes from year to year. An unlimited number of students may work on the project. However, only five students may present the finished project at the competition.

Awards: Scholarships and prizes.

For Information Write To:

> **JETS**
> 1420 King Street #405
> Alexandria, VA 22314

Web Site: http://www.asee.org/jets/teams

Junior Science and Humanities Symposium (JSHS)

Participants: Students in grades nine through twelve.

Format:

Round I. Regional. Students submit completed independent research papers to regional coordinators. Selected papers are then presented to judges. The forty-five regional symposia are conducted throughout the

United States, Puerto Rico, and various United States Army bases around the world.

Round II. National. Students who placed first at each of the regional competitions compete in the national Junior Science and Humanities Symposium. The alternates also attend the presentations.

Awards:

Round I. Regional. One finalist and four alternates are selected to attend the National Junior Science and Humanities Symposium. Only the finalists will present and compete at the national competition.

Round II. National. Scholarships and an expense-paid trip to the London International Science Youth Forum.

For Information Write To:

> **Academy of Applied Science**
> 98 Washington Street
> Concord, NH 03301

Web Site: http://www.jshs.org

NASA Student Involvement Program (NSIP)

Participants: Students in grades seven through twelve.

Format: The program provides opportunities for students and teachers to expand their knowledge of science, mathematics, engineering, technology, and geography through competitions that include actual or simulated problems or challenges of "real world" science, mathematics, engineering, technology, and geography activities.

Awards: Certificates, plaques, and all-expense paid trips.

For Information Write To:

> **NASA Student Involvement Program**
> Technical Education Research Centers
> Center for Earth and Space Science Education
> 2067 Massachusetts Avenue
> Cambridge, MA 02140

Web Site: http://www.nsip.net

Science Olympiad

Participants: Students in grades nine through twelve.

Format:

Round I. Regional. The Science Olympiad is a team competition. A team is composed of fifteen students who participate in over twenty different events. The events change every year and are chosen from all areas of science including biology, chemistry, Earth science, and physics as well as mathematics and computer programming.

Round II. State. Winners of the regional events are eligible to compete in the state event.

Round III. National. Winners of the state event are eligible to compete in the national competition.

Awards: Individual gold, silver, and bronze medals are awarded for each event at each level. Team trophies are also awarded. Scholarships are awarded at the national competition.

For Information Write To:

> **Science Olympiad**
> 5955 Little Pine Road
> Rochester, MI 48306

Web Site: http://www.macomb.k12.mi.us/science/olympiad. htm

State Science Talent Searches

Participants: Students in their last year of high school.

Format: The state Science Talent Search bases its competition on the research papers of graduating high school students who are also entering the national Science Talent Search. The searches are held concurrently with the national Science Talent Search by special arrangement with Science Service.

Awards: Certificates and cash awards.

For Information Write To:

> **Science Service**
> 1719 N Street, NW
> Washington, DC 20036

Web Site: http://www.sciserv.org

Science Talent Search

Participants: Senior high school students.

Format:

Round I. The Science Talent Search is a competition to discover seniors in high school who have the potential to become the research scientists and engineers of the future. Students submit a paper describing an original science research project and the results. The paper should be a minimum of five pages and a maximum of twenty pages in length.

Students also complete an entry form containing objective questions about the student, the research project, and the student's promise as a scientist. Students complete several essays on topics such as "scientific attitude" and "scientific curiosity." Students submit copies of their transcripts, SAT scores, and letters of recommendation from teachers and mentors.

Round II. Interviews with the finalists.

Awards: Semifinalists receive certificates and national recognition. Finalists receive an all-expense paid trip to Washington, D.C., and scholarships.

Sponsor: The STS is administered by Science Service and funded by the INTEL Corporation.

For Information Write To:

> **Science Service**
> 1719 N Street, NW
> Washington, DC 20036

Web Site: http://www.sciserv.org

Siemens Westinghouse Science and Technology Competition

Participants: High school students. Students may enter either individually or in teams of up to three students.

Format:

Round I. Six geographic regional competitions.

Round II. National competition for regional finalists. Interviews with finalists.

Awards: Over one million dollars in total awards annually. Medals and awards to regional winners and semifinalists. Internship opportunities for successful participants. National champion award.

For Information Write To:

Siemens Foundation
1301 Avenue of the Americas
New York, NY 10019

Web Site: http://www.siemens-foundation.org

ThinkQuest

Participants: Students in grades seven through twelve participating in teams of two or three.

Format: Students produce sets of Web pages as teaching and learning tools for use by teacher and students around the world. Projects may be entered in five awards categories: arts and literature, interdisciplinary, science and mathematics, social sciences, and sports and health.

Awards: Over a million dollars in scholarships and awards.

Web Site: http://www.advanced.org

Appendix B

Titles of Winning Projects

Titles of projects that have been successfully entered into either the Science Talent Search or the International Science and Engineering Fair are listed below.

They are listed in thirteen project categories that are often found in the International Science and Engineering Fair. The selected categories are behavioral and social sciences, biochemistry, botany, chemistry, computer science, Earth and space science, engineering, environmental science, mathematics, medicine and health, microbiology, physics, and zoology.

Behavioral and Social Sciences

Behavioral and social sciences includes psychology, sociology, anthropology, archaeology, ethnology, linguistics, animal behavior (learned or instinctive), learning, perception, urban problems, reading problems, public opinion surveys, educational testing, etc.

Sample project titles

- Anatomical and Cognitive Features of Preattentive Visual Processing Through Psychophysical Experimentation

- Testing the Effect of Suggestions on Subliminal Perceptions

- A Study for the Preference for Peer Counseling Among Children of Divorce

- Relative State Power in the United States Senate

- Statistical Survey of Perceptions of United States Financial Spending at Ward Melville High School

- Myers-Briggs Psychological Types and Scholastic Achievement

- *Odin* as an Indefinite Article in Russian and the Universal Evolution of the Numeral "One" into an Indefinite Article

- A Survey of Geographical Knowledge

- Gender Differences in Everyday Knowledge

- Chess Ability: A Study into Linguistic Ability Versus Intelligence Quota

- A Study of the Possible Relationships Between Grade Point Average, Sex, and Logic Testing Technique and Ability

- The Effects of Media on Political Campaigns

- The Comparison of the Proportionality of Left-Handed People in Honors and Accelerated Classes as Compared to that in Regular Classes

Biochemistry

Biochemistry includes molecular biology, molecular genetics, enzymes, photosynthesis, blood chemistry, food chemistry, hormones, etc.

Sample project titles

- Movement of Neutrophils Stimulated by Chemoattractant

- The Effects of Ethylene Glycol on Conductivity and Diffusion in the Septate Median Giant Axon of the Earthworm *Lumbricus terrestris*

- Analysis of Assumed Correlation Between the Concentration of Ascorbic Acid in Fruit Juices and Human Taste Response

- Weak Inhibition of Activated Human Coagulation Factor X by Soybean Trypsin Inhibitor (Kunitz)

- Double NOR (dNOR) as a Risk Factor for Trysomic Offspring

- Comparing the Effects of Brain Extract and Ascorbic Acid on Acetylcholine Receptor in Chick Myotubes

Botany

Botany includes agriculture, agronomy, horticulture, forestry, plant biorhythms, paleontology, plant anatomy, plant taxonomy, plant physiology, plant genetics, hydroponics, algology, mycology, etc.

Sample project titles

- Tomato Spotted Wilt Virus

- The Effects of the Algae Codium Fragile Used as a Fertilizer on the Growth of Plants: A Two-Year Study

- The Effect of Phosphorus and Magnesium on Radish Plants

- The Effects of Acid on Plant Growth

Chemistry

Chemistry includes physical chemistry, organic chemistry (other than biochemistry), inorganic chemistry, materials, plastics, fuels, pesticides, metallurgy, soil chemistry, etc.

Sample project titles

- Organic Chemistry: Formal Replacement of Oxygen by C_5H_4 Groups to Form Organic Carbonate

- The Use of Synthetic Protein as a Substitute for Phytoplankton with Rotifers

- A Transient Electric Birefringence Study of DNA Fragments in Agarose Gels

- Boundary Layer in Viscous Fluid Flow

- The Liberation of Hydrogen from Water

- Volatile Chemical Utilization for Domestic Energy Cooling

- Ionization in Water's Surface

- Theory Analysis of Chloride Induced Stress Corrosion Cracking in Austenitic Stainless Steels

Computer Science

Computer science includes new developments in software or hardware, information systems, computer systems organization, computer methodologies, and data (including structures, encryption, coding and information theory), etc. (Projects that use existing computers or apply computer procedures to scientific problems should be entered in the area of basic science to which the application is made, not in the computer science category.)

Sample project titles

- Computer Methodologies: Statistical Analysis of Randomness of Random Number Generators

- Interactive Human Networking via Interprocess Communication (the CHAT System)

- TOOLS—A Natural Language Database System

- Computer Simulated Behavior of a Network of Model Neurons

- Development of a Meta-Language Computer Programming Environment

- General Purpose Anti-Virus Program in C

- Distributed Parallel Processing

Earth and Space Science

Earth and space science includes geology, geophysics, physical oceanography, meteorology, atmospheric physics, seismology, petroleum, geography, mineralogy, topography, optical astronomy, radio astronomy, astrophysics, etc.

Sample project titles

- Control of Growth Patterns: The Response of Bean Roots to Centrifugal Force and Gravity

- Fundamental Design Concepts of a Low-Tech, Cost-Effective Space Transportation System

- On the Trochoidal Waves of Gerstner and Rankine

Engineering

Engineering includes civil, mechanical, aeronautical, chemical, electrical, photographic, sound, automotive, marine, heating and refrigerating, transportation, and environmental engineering. It also includes power transmission and generation, electronics, communications, architecture, bioengineering, lasers, etc.

Sample project titles

- Mechanical and Electrical Feedback: The Eyes, Ears, and Senses of Robots

- Demonstrations of Single-Beam Holography and the Various Conditions Causing Vibration

- The Carnot-Diesel Cycle: An Ideal for Internal-Combustion Engines?

- Enhancement of Building Heating System: A Measure of National Energy Conservation

- Light Feet—Electro-Magnetic Sneaker

- Heat Convection Mechanism of Envelope Houses

- Using a Vertically Rotating Propeller to Increase the Efficiency of the Traditional Windmill

- Efficient Methods of Conserving Energy Through the Insulation of Basements

- Pressure Activated Walkway Lights Using Relay Switches

- The Reallocation of Waste Heat in Water Disposal Systems

- Finding the Most Efficient Blade Angles for a Dutch Windmill

- A Comparison of Various Modern Wind Turbine Variables with Their Possible Effectiveness on Primitive Windmills

Environmental Science

Environmental science includes pollution (air, water, land), pollution sources and their control, waste disposal, impact studies, environmental alteration (heat, light, irrigation, erosion, ecology, etc.

Sample project titles

- The Practical Use of Earth Sheltered Housing in Energy Conservation

- Fires' Role in the Natural Ecosystem with Specific Investigation and Recommendation for Land Management Policies in Suffolk County, New York

- High Efficiency Hydro-Vaporizing Cooling System

- Reducing Heat Loss in Solar Homes by Applying Synthetic Materials

- Exothermic Reaction as an Alternative Energy Source for Heating System

- The Compost Water Heater

- Methane Energy from Bacterial Decay of Garbage

- Water Purification Through an Enhanced Evaporation-Condensation Process

Mathematics

Mathematics includes calculus, geometry, abstract algebra, number theory, statistics, complex analysis, probability, topology, logic, operations research, and other topics in pure and applied mathematics.

Sample project titles

- Plotting Catenaries and Other Hanging Chain Curves Using Vector Constructions

- Finite Polynomial Fields in Binary

- Fourier Polynomials and Polyhedrons

- Conjectures Towards a Polynomial Expansion Theorem

- Application of the Laplace Transform to the Solutions of Certain Initial Value Problems of Differential Equations

- Cubic Spline Interpolation

- Mathematical Design of an Efficient Collector of Wave Energy

- Hysteresis and the Structure of Fractal Basin Boundaries in the Hénon System

- Factoring (X^{n-1})

- Toward a General Definition of Chaos

- On Constructing Polygons on Orthogonal Integral Lattices

- Comparing Cardinalities of Infinite Sets

- A Methodology for Projecting the Population Center of the United States

Medicine and Health

Medicine and health includes medicine, dentistry, pharmacology, veterinary medicine, pathology, ophthalmology, nutrition, sanitation, pediatrics, dermatology, allergies, speech and hearing, optometry, etc.

Sample project titles

- Determination of the Mechanism of Anesthetic Action Through the Electro-Neurophysiological Analysis of the Temperature Dependence of Various Kinetic Properties in BC3H-1 Murine Intracranial Tumors

- Lyme Borreliosis in Cattle: Characterization of the Immune Response and Evaluation of Serological Testing

- The Contribution of the Neutralization Antigenic Site I to the Immunogenicity of the Poliovirus in Human Beings

- Characterization of the Generation and Inhibition of Factor Xa in the Human Blood Coagulation System

Microbiology

Microbiology includes bacteriology, virology, protozoology, fungal and bacterial genetics, yeast, etc.

Sample project titles

- Development of a Method Using Household Reagents to Deproteinize DNA Extracted from *Artemia salina*

- ELISA Analysis of Some Complement Proteins

Physics

Physics includes solid state, particle, nuclear, atomic, and plasma physics, optics, acoustics, superconductivity, fluid and gas dynamics, thermodynamics, semiconductors, magnetism, quantum mechanics, biophysics, etc.

Sample project titles

- Diode Laser Narrowing Using a Diffraction Grating

- The Effects of Colorants and Temperature on a White Clay Body

- An Experiment to Evaluate the Validity of the Yttrium Barium Copper Oxide-Based Superconductor

- Creating Resonance in Rubidium-85 Using an AlGaAs-GaAs Diode Laser

- Piezo-Electric Nighttime Illumination System

- Fiber Optics: Increasing Lighting Efficiency

Zoology

Zoology includes animal genetics, ornithology, ichthyology, herpetology, entomology, animal ecology, anatomy, paleontology, cellular physiology, animal biorhythms, animal husbandry, cytology, histology, animal physiology, neurophysiology, invertebrate biology, etc.

Sample project titles

- Exogenous ATP Is Not Required for Pronuclear Formation in Vitro

- A Study of the Correlation Between Left and Right Clawed Fiddler Crabs and Their Body Dimensions

- Observing Two Competing, Mating Males of the Firefly, *Photinus macdermotti*

- Effects of Various pH Levels on the Growth Patterns of *Artemia salina*

- Discovering Female Response to Altered Male Flash Patterns in the Photinus Firefly Mating System

This list has given you an indication of the kinds of projects and experiments completed by high school students for science research competitions. However, you may feel that you would like even more information about specific projects and the kinds of papers actually presented at the International Science and Engineering Fair.

Each year, Science Service prints a book of abstracts of the finalists' projects in the International Science and Engineering Fair.

For Information Write To:

Science Service, Inc.
1719 N Street, NW
Washington, DC 20036

Appendix C

Summer Student Science Training Programs

Working on a research program during the summer or after school at a camp or at a university can be quite exciting. As you work with professional scientists, you will discover that you will be treated as a professional. Your opinions and your work will become important. You will find that the work you complete will become an intrinsic part of the project.

Summer Programs

There are many student science training programs available to high school students interested in working on science research projects. Programs are available in the summer or throughout the year.

One of the easiest ways to get started in science research is to attend a summer student science training program. These programs can be either day or residential camps. Quite often positions in the camps are filled on a competitive basis.

There are three major categories of camps: (1) Students are taught course work about various topics in order to gain background information for future research projects. (2) Students are taught course work and at the same time provided with a mentor in order to become involved in independent research for part of each day. (3) Students are set up in a laboratory immediately. They work throughout the summer on a research project.

All of these programs usually assign each student to a mentor and help the student develop a set of simple original experiments for a science research project.

In most cases the students are expected to pay for room, board, travel, and tuition. However, scholarship money is available for many students.

College Programs

During the summer months many colleges allow students to participate in college-level courses. These courses not only give students a taste of the scope and level of a college course but can be directly related to a research project.

After signing up for the course, speak to the professor or department chair about your goal of working on a science research project. Many times the professors will allow students to observe the projects and experiments in progress and to participate in some research work in the lab. The colleges will also allow students to use the library facilities for their research.

Courses for Talented Youth: CTY

Qualified students with a very high score on the SAT taken before their thirteenth birthday will be able to attend courses for talented and gifted students at several different sites throughout the country. The score required for entry into the

Johns Hopkins CTY program varies with age. For specific information about the required score, write to CTY.

Although these courses are not research-oriented, they will give students an excellent background in many science and math areas. These courses may give students ideas on topics that could be used for science research projects.

For Information Write To:

Center for Advancement of Academically Talented Youth (CTY)
The Johns Hopkins University
Charles and 34th Street
Baltimore, MD 21218

Web Site: http://www.jhu.edu/~gifted/programs.html

Appendix D

Summary of Science Research Competition Rules

Each competition that you enter will have a set of specific rules and guidelines for you to follow. The first thing you must do whenever you enter a contest is to read all the rules and regulations thoroughly. I always suggest that you read them through twice and take notes. Place this information in your journal. Then as you work on your project or experiment, follow the rules of your competition exactly. It is important to review the rules from time to time as you work on your project. Sometimes you may have forgotten the exact format of the rules. Or you may have changed your project slightly and may find that your work is not in accord with the rules of the competition you are entering.

Because the two largest science research competitions are the Science Talent Search and the International Science and

Engineering Fair, here is a summary of the rules for each of these competitions.

For a complete set of rules for both the Science Talent Search and the International Science and Engineering Fair, visit the Web site <http://www.sciserv.org>, or write to:

Science Service
1719 N Street, NW
Washington, DC 20036

Science Talent Search

Each contestant in the Science Talent Search is asked to submit evidence of his or her fitness for the further study of science in the following ways:

1. By writing a report on an independent research project in the physical sciences, behavioral and social sciences, engineering, mathematics, or biological sciences (excluding live vertebrate experimentation).

2. By submitting a completed entry form.

The Science Research Report

Each student must write a report on his or her science research project. It must be the composition of the student and should be creative, original, and interpretive in character. It should tell what the student is doing or has done in science experimentation or other research activity and be written in the manner of a scientific paper. It should be a minimum of five pages and a maximum of twenty pages in length, typed, double-spaced on white paper, on one side only, and stapled in the upper left-hand corner. The full name of the student should be on the upper right-hand corner of each sheet.

In addition, each student *must* include with the report a seventy-five-word summary of the project in language that can be understood by nonscientists. This should follow the title page.

The report should stand by itself. Adding an appendix to include material that exceeds the work limit is not as good as finding a way to summarize it and stay within the word limit. The judges are very impressed with quality—not quantity.

Entry Form

The entry form is designed to furnish information about the student and his or her academic training. There are four parts to the entry form, and each part must be filled out carefully.

Eligibility

1. The Science Talent Search is open to any student who is in the last year of secondary school.

2. Students entering the competition are expected to complete college entrance qualifications before October 1 of year of graduation, and must not have competed in any previous Science Talent Search.

3. The immediate families of employees of Science Service, Inc., judges and evaluators of the Science Talent Search, and employees of the sponsoring company, their affiliates, and subsidiaries are not eligible to enter the Science Talent Search.

4. There are no age limits.

Requirements

1. Each student must submit a report on an independent research project.

2. Each entry must include the student's high school transcript and an entry form filled out by the student, his or her teachers, and principal.

3. The high school transcript *must* include available standardized test scores, e.g., ACT, NEDT, PSAT/NMSQT, SAT.

4. No projects involving live vertebrate experimentation will be eligible. However, if a student is working in a laboratory where animal experimentation is taking place, the student's research is eligible for entry in the Science Talent Search (1) if the student has no physical contact with the animals; (2) if the material on which the student is working (tissue, blood, etc.) is supplied to the student by the supervising scientist; and (3) if the animals involved are sacrificed for some purpose other than the research being done by the student.

5. Projects involving behavioral observations of animals in their natural habitat are excluded from the above ruling and are eligible.

6. Use of human subjects: All proposed human contact studies, including surveys and questionnaires, must be evaluated for risk (psychological and physical) by an institutional review board, and reports on research involving human subjects must be accompanied with approval by an institutional review board and an appropriate informed consent statement. These forms can be found in the Science Talent Search entry form.

As a general policy, students should be circumspect when undertaking research projects involving human subjects in either behavioral or biomedical studies. The following is a legal definition:

Human subject is a person about whom an investigator (professional or student) conducting scientific research obtains (1) data through intervention or interaction with the person or (2) identifiable private information.

7. The research project *must* be the work of a single individual. Group projects are not eligible.

8. A student who has attended a precollege student research program and has used material from that experience as part of his or her project *must* submit a statement from the supervising scientist of that program, giving descriptive and explanatory evidence of independence and creativity to Science Service before the deadline.

9. The Science Talent Search entry deadline is in early December and varies from year to year.

International Science and Engineering Fair

A summary of the rules for participants in regional ISEF affiliated fairs and in the ISEF follows. Each of the affiliated ISEF fairs follows the exact rules of the ISEF furnished by Science Service. Be sure to check current rules, since ISEF makes several minor rule changes each year.

1. Students may compete only within the region of the ISEF in which their school is located.

2. Entrants must bring an original research plan to the ISEF. The research plan is slightly different from a research paper. Although many of the sections and section titles are the same, the emphasis of what to include in each section is quite different.

The research plan includes

   a. **Problem and hypothesis.** The student should present in this section the ideas behind the research. Why is the research to be done? There should be a reason for wanting to do the research.

   b. **Methods and procedures.** Methods to be used in the research should be presented in detail. For example, if an animal or embryo is to be treated with drugs, the name of the drug, the concentration and amount of the drug to be administered, and the method of administration should be clearly outlined.

   c. **Bibliography**

3. Not more than two students and one team may be selected to represent a regional fair at the ISEF. They *must* be students enrolled in the ninth, tenth, eleventh, or twelfth grade in a public, private, or parochial school.

4. Exhibit size is limited to 76 cm (30 in) deep, front to back; 122 cm (48 in) wide, side to side; and 274 cm (108 in) high, floor to top. (Tables are 76 cm high.)

5. Two copies of a typed abstract about the student's research, including purpose, procedures used, results (data), and conclusions are required. The abstract should not include acknowledgments, work, or procedures done by the mentor. Only minimal reference to previous work may be included. The abstract must be on the official ISEF abstract form. The abstract should be no longer than two hundred fifty words.

6. Research plan and all approval forms must be included in the display. Check the current guidelines for all necessary forms.

7. All necessary forms can be found at <http://www.sciserv.org>.

Appendix E

The Science Talent Search Entry Form

The Science Talent Search entry form is a very important part of the entire Science Talent Search. It consists of four different parts. Included in Part I are many simple informational questions for you to answer. In Part II, there are five short essay questions. Each of the essays is about a different scientific trait that will describe to the judges your potential to become a creative scientist. The judges will be looking for very specific examples about your accomplishments for each of the scientific traits. In Part III, your teachers are asked to describe their observations and inferences of you and your work that will help the judges predict whether you will actually become a creative scientist. Last of all, in Part IV, your principal and guidance department are asked to describe your scholastic achievements. This will include your rank in class and a standardized test score report.

In addition, there are several supplemental forms to be filled out for research projects involving human subjects and vertebrate animals.

Part I

Part I consists of simple informational questions that are used to describe your general background. You will be asked to list such various items as your name, address, date of birth, and project title. In addition, there are several questions about your likes and dislikes in school as well as lists of your high school awards, honors, and hobbies.

There is one important short essay in this section in which you will be asked to describe what you have learned from your Science Talent Search report.

An excellent example is from a winning project, "Toward a General Definition of Chaos."

> I discovered what it meant to be a working scientist, and learned yet again how much fun it is to explore the unknown. This was the first time I had the opportunity to conduct cutting-edge research in any field, and my excitement at searching for and developing new ideas was tremendous. I also learned a great deal about chaos theory, an area I had been interested in but only marginally familiar with beforehand. My research project provided an opportunity for honing my ad hoc programming skills and developing problem-solving strategies based on numerical analysis. I confirmed my long-held belief that I want to be a scientist, and I feel blessed to be able to pursue a career doing what I enjoy best.

Part II

In this section, you will be asked to answer five essay questions about your scientific traits or characteristics that may help predict your future as a creative scientist. The five traits are scientific attitude, curiosity, inventiveness, initiative, and work habits. A score is given for each trait or "trait family" for which concrete and objective evidence is given. In other words, you must state specifically what you have done that demonstrates competency in that trait. A general or vague answer will receive no points from the judges. The judges are looking for very specific and concrete examples of your qualities and accomplishments that reflect each of these traits. You should take your examples from your current research as well as any other projects and research that you have ever worked on. You can also include examples from your classes, activities, hobbies, and daily life.

Scientific Attitude

The question states, What have you done that demonstrates your scientific attitude? Do you discriminate between pertinent and nonpertinent evidence in solving a problem? Do you "try it and see"? What is your approach to solving a problem? Specific examples should be provided.

A good example can be found in a winning project, "Hysteresis and the Structure of Fractal Basin Boundaries in the Hénon System."

> I have taken several systems of quadratic equations, including the Hénon, and scientifically investigated the exact effect of parameters on the outcome (basins, etc.). I then reversed my approach. I tried to find appropriate parameters for certain results I expected. I was able to do this after having gathered immense amounts of experimental evidence (by "try-and-see") of which I used only the pertinent results but discarded none "for future reference." Thus I was able to acquire a deeper understanding of nonlinear dynamics and gain new insights into such a well-researched system as the Hénon. For the STS report, I began with experimental research which enabled me to discover basic theorems, and I could then prove them rigorously.

Curiosity

The question states, What ideas or devices have particularly attracted your attention and what have you done about them? A good example can be found in a winning project, "Toward a General Definition of Chaos."

> I have always been fascinated with the beauty and symmetry of nature. In combination with my love for mathematics, this appreciation has led me to seek mathematical descriptions that capture some facets of nature's mystery. I was thus drawn early on to a study of Fibonacci series and the Golden Section, which are manifested in a myriad of seemingly unrelated natural phenomena. In the same way, I embarked on a study of fractals, working through Mandelbrot's classic books to understand these microcosmic imitations of nature. It is exciting to me that several simple conditions can produce a complex pattern. I have studied the Hénon map and various other iterated relations that produce strange attractors, in order to learn more about the phenomenon known as chaos. In addition, I researched chaotic techniques as an alternate method for analysis of periodic oceanographic time series. This past summer, I studied the Duffing Oscillator, a chaotically behaving model of an electric circuit. I wondered what the distribution of frequencies in the model's oscillations would be, and began development of a measure of "chaoticity" based on frequency analysis.

Inventiveness

The question states, Have you ever made a discovery that was exiting and new to you? How did it come about and what was it? Have you ever produced an idea or experiment that was new to everyone? Describe it and the circumstances surrounding the discovery. Specific examples should be provided to ensure proper evaluation of your promise as a scientist.

An excellent example is Steven's. Although his Science Talent Search project was titled, "Diode Laser Narrowing Using Diffraction Grating," Steven chose to write about an invention he created for the New York State Student Energy Research Competition. Steven won first prize in New York State with his invention.

> An example of my inventiveness was the design of an air-conditioner with an efficiency rating higher than conventional air-conditioners. I wanted to find a system that was easy to install, had a low power consumption, and no negative environmental effects. My first design was an improvement over the humidifying cooler used in the southwest. But because of the high humidity in the northeast, its performance was not satisfactory. So I designed a second system. Only this time, I used an endothermic chemical reaction using non-toxic chemicals. This design was much more effective since it was not dependent on humidity. I also designed a thermostat network that optimized the efficiency of the cooler.

Initiative

The question states, What have you done outside of your school curriculum that shows initiative? How did you pursue these activities, what difficulties did you encounter and how did you handle them? Specific examples should be provided to enable a proper evaluation of your promise as a scientist.

An excellent example can be found in David's winning Science Talent Search entry that accompanied his paper titled, "Hysteresis and the Structure of Fractal Basin Boundaries in the Hénon System."

> I started mathematics on my own starting in junior high school. I learned the equivalent of 1st and 2nd year college calculus, linear algebra, vector analysis, complex analysis, and set theory. In addition, I studied the history and philosophy of math, computer languages, music and art appreciation and even more. I picked up

the relevant books and read them on my own. I encountered difficulties only when I jumped nonsequentially to more advanced material. In that case I looked up the problematic developments elsewhere or gathered the meaning from the context. Chaos and fractal theory was another important topic I studied independently. I looked up and read EVERYTHING! (I kept a journal of all these articles.) I also followed the development of new and important findings directly through professional articles. I searched for relevant articles in all the important math journals and began investigating the system "for my own enlightenment."

Work Habits

The question states, Do you plan your work in advance? Are you more interested in the overall planning of a project, or in the specific details? Do you complete your work on time? Do you allow sufficient time for your work, or do you try to do most of it "at the last minute"? Specific examples should be provided to enable a proper evaluation of your promise as a scientist.

One student wrote the following about his work habits for his winning STS report:

> In general, I create a work plan in advance and begin by following it. However, I then tend to follow any leads towards the most promising prospect, adapting my plans to evidence and circumstance. Therefore, while working, I tend to concentrate on immediate details but once I've reached a goal I consolidate the overall structure based on it. Thus I did in my JSHS paper, which I began by an investigation of the Parameter Plane Mandelbrot like structure of the Hénon system. This led me to discover interesting aspects of the mapping plane basins, which became the core of my research, finally integrating all the results into "Parametric Evolution of Fractal Basins." In the current paper, I began by investigating hypotheses describing the system. Then, as

the deadline approached I suddenly discovered that these relations can be explained fully using basic properties of manifolds. Thus my project changed completely in the course of two weeks.

In Part II, you will also be asked to describe how you got the idea for your research, where your research was done, who supervised your research (if anyone), and what help you received in doing your research (if any).

If you did your research in a scientist's laboratory and used material from that experience as part of your research, the scientist with whom you worked must send to Science Service a letter answering the following questions:

1. How did the student get the idea for the research?

2. Did the student work on the research as part of a team or group?

3. How independently did the student work on the research?

4. What did the student do that showed creativity and ingenuity?

5. What was the intensity of the program?

6. Has the student received a salary or other compensation for doing his or her research?

This is a very important letter. If you worked with a supervising scientist, he or she can either send the letter directly to Science Service or have the letter included with your application. However, if the letter is not received by Science Service prior to the deadline, your application will be considered incomplete and not judged for the Honors or Finalists Groups.

Part III

Part III of the entry form should be filled out by the teacher or advisor who supervised your research. You can have more than one teacher fill out this part. Your teacher will be asked to answer the following questions:

1. How long have you known this student and in what capacity?

2. Please summarize fully your observations and inferences that would be useful to the judges in predicting whether the student will become a creative scientist. Relevant topics here include (but are not restricted to): scientific attitude, curiosity, initiative, originality of thought, and work habits.

3. What are the student's strengths in areas such as peer relationships, outside activities, and family life?

4. What do you consider the student's principal weakness?

5. It would be most helpful in our considering this student if you will answer carefully the following question: What aspects of this student's abilities, work habits, or overall functioning might present difficulties for him or her?

6. To what extent is the research the work of the student? Did he or she do all that is claimed in the research report?

It is a good idea to go over these questions with your teacher or advisor so that your teacher will be able to fill out Part III knowledgeably.

Part IV

In this section of the entry form your principal or guidance counselor will describe your scholastic achievements. This will include your rank in class and a standardized test score report. You must include an official transcript with your completed Science Talent Search application.

Signatures

Last of all, you must validate your application. You, your advisor, and principal are asked to sign your entry form to certify the truth of your essays, statements, and scores.

As you can see, this application is difficult and complex to complete. However, it is a vital part of the judging in the Science Talent Search. So take your time and work very carefully and thoughtfully on the entry form.

Internet Addresses

The following is a summary of the national competitions and their Web sites. Since contests change over time, you can check for further information by looking for "Science Contests" and "Math Contests" on various search engines.

Dupont Challenge/Science Essay Competition
http://www.glcomm.com/dupont

Duracell Scholarship Competition
http://www.nsta.org/programs/duracell.shtml

International Model Bridge Building Competition
http://www.iit.edu/~hsbridge/

International Science and Engineering Fair
http://www.sciserv.org

International Science Olympiads in Mathematics, Physics, Chemistry, Biology, Computer Science, and Astronomy
http://www.olympiads.win.tue.nl

Junior Engineering and Technical Society
http://www.asee.org/jets/teams

NASA Student Involvement Program
http://www.nsip.net

Science Olympiad
http://www.macomb.k12.mi.us/science/olympiad.htm

Science Talent Search
http://www.sciserv.org

Siemens Westinghouse Science and Technology Competition
http://www.siemens-foundation.org

State Science Talent Searches
http://www.sciserv.org

ThinkQuest
http://www.advanced.org

Index